National Science and Technology Council

Committee on Science

Fast-Track Action Committee on Research and Development Reporting Standards[1]

Federal R&D Reporting Model

August 11, 2016

---

[1] FTAC-RDRS participants listed in Appendix A.

Contents

Purpose

The Fast-Track Action Committee on Research and Development Reporting Standards (FTAC-RDRS) was established to provide Federal agencies with tools and guidance to address consistency and accuracy of research and experimental development (R&D) funding data provided by agencies in response to several Federal-wide data calls. It also kick started a long-term Federal R&D Community-of-Practice to continue to address questions and concerns that could not be addressed by the FTAC-RDRS.

Introduction

Virtually every analysis of U.S. scientific and technical activity uses as its foundation the data on Federal research and development (R&D) spending collected through the National Science Foundation (NSF) and the Office of Management and Budget (OMB) surveys of Federal agencies. The data are used by government, academia, industry, and a host of nonprofit analytical and advocacy groups as the primary source of information about Federal spending on R&D. These data are also used by Federal agencies, the White House, and Congress as the basis for making budgetary and policy decisions about the Federal R&D enterprise. As a recent National Academies (2010) report states, the data are "used to reach conclusions about important and fundamental policy questions, such as whether a given field of research is being adequately funded, whether funding is balanced among fields, whether deficiencies in funding may be contributing to a loss of U.S. scientific competitiveness, and which agencies are most important for the health of a scientific discipline."[2]

For all of these reasons, it is important for these data to be accurate, timely, high quality, and to be collected as efficiently as possible.

Nonetheless, there are longstanding concerns about the consistency, accuracy, and quality of R&D spending data reported to both OMB and the NSF surveys. Despite the same definition of R&D, data reported to the OMB and NSF data calls differ even when adjusted for comparable basis. The FTAC-RDRS was formed to address these concerns and to help Federal agencies make improvements in both their R&D spending data and the processes used to identify, collect, and report the data. At the conclusion of its work, the FTAC co-chairs believe the FTAC has made significant progress in identifying measures to improve both the consistency, accuracy, and quality of R&D spending data and streamlining agency processes for collecting and reporting the data.

Government-wide Definitions of R&D

The first goal in the FTAC-RDRS charter was to "collect and review existing R&D definitions and reporting requirements, identifying opportunities for clarification and simplification that are consistent with the internationally comparative standards."[3]

---

[2] National Research Council, 2010, *Data on Federal Research and Development Investments: A Pathway to Modernization*. pg. 7.
[3] See FTAC-RDRS Charter, pg. 2.
https://community.max.gov/download/attachments/913670779/FTAC%20RDRS%202015%20charter%20-%20signed.pdf?version=1&modificationDate=1450292907483&api=v2

To address this goal the FTAC-RDRS compiled all existing Federal-wide data calls and definitions. Since all of these data calls and definitions reside in separate documents and locations, a list of these sources with corresponding web links was posted to the FTAC-RDRS MAX page, Documentation Archive, under the title of *Supporting Materials*:

https://community.max.gov/pages/viewpage.action?pageId=1022788437.

R&D is defined similarly in:

- OMB Circular A-11, Schedule C
- OMB Circular A-136, Section II.4.10
- Statement of Federal Financial Accounting Standards 8, Chapter 7
- Survey of Federal Funds for Research and Development, NSF

Additional resources referencing R&D definitions in the Federal Acquisition Regulations and the Department of Defense (DOD) Research, Development, Test, and Evaluation (RDT&E) Budget Activities along with the items above are consolidated in a draft document by the National Science Foundation, also posted to the MAX page. In general, the government-wide definitions of R&D are intended to be consistent with the historical and international definitions governed by the Organisation for Economic Cooperation and Development's (OECD).

Glossary of Terms Relevant to Government-wide R&D Reporting[4]

FTAC-RDRS Members also requested that a Glossary of Terms relevant to federal R&D reporting would be valuable for future reference. We have included the glossary here below. The definition of R&D provided below is the same used in both OMB Circular A-11 and the *Survey of Federal Funders for Research and Development*, and conforms to the international standard described in the OECD's Frascati Manual *2015: Guidelines for Collecting and Reporting Data on Research and Experimental Development*.

Term/Abbreviation	Definition
Applied Research	Original investigation undertaken in order to acquire new knowledge; it is, however, directed primarily towards a specific, practical aim or objective.
Basic Research	Experimental or theoretical work undertaken primarily to acquire new knowledge of the underlying foundations of phenomena and observable facts, without any particular application or use in view.
BERD	Business enterprise Expenditure on R&D represents the component Gross domestic Expenditure on R&D (GERD, see below) incurred by units belonging to the business enterprise sector.
Budget Authority	The authority provided by law to incur financial obligations that will result in outlays.[5]
Budget Function	The Federal budget is divided into 20 categories known as functions. These functions include all spending for a given topic without regard to the Federal agency that oversees individual Federal programs related to these functions. Both the President's

[4] Sources include the OECD Frascati Manual (2015); OMB Circular A-11, Schedule C; OMB Circular A-136, §II.4.10; Statement of Federal Financial Accounting Standard 8, CH. 7; NSF Survey of Federal Funds for Research and Development; OMB Statistical Policy Directive 16 (1978); and others.
[5] https://www.whitehouse.gov/sites/default/files/omb/assets/a11_current_year/s20.pdf

	Budget and Congressional Budget Office reports use these 20 functions, such as National Defense and Health.[6]
Character of Work	See Type of R&D
Experimental Development	Systematic work, drawing on knowledge gained from research and practical experience and producing additional knowledge, which is directed to producing new products or processes or to improving existing products or processes.
External R&D Funds	Funds spent on R&D that originate outside the control of the reporting unit.
External R&D Personnel	Independent (self-employed) or dependent (employees) workers fully integrated into a reporting unit's R&D projects without formally being persons employed by the same R&D-performing unit.
Extramural R&D	All R&D performed outside of the reporting units. Funds for extramural R&D should include only internal funds (not from external sources) provided to an outside unit for R&D performance, including either where there is an expected compensatory delivery of R&D (e.g., contract), or where no compensatory delivery is expected (e.g., grant). Funds for extramural R&D often include payments for costs other than for core R&D activities, such as cost elements covering depreciation costs, performer profit, delivery charges, etc.
FFRDC	Federally Funded Research and Development Centers are privately-operated R&D organizations that are exclusively or substantially financed by the Federal government in order to meet some special long-term research or development need which cannot be met as effectively by existing Federal government in-house or contractor resources.[7]
FFS	Federal Funds Survey, formally known as the NSF Survey of Federal Funds for Research and Development.[8]
FLC	Federal Laboratory Consortium for Technology Transfer[9]. This statutorily created consortium of Federal laboratories is supported by a percentage set-aside of Federal agencies' intramural R&D funds.
FM	Frascati Manual – Guidelines for Collecting and Reporting Data on Research and Experimental Development is the international consensus document published by the OECD for defining and reporting R&D.[10]
FORD	Fields of Research and Development – previously referred to as fields of Science and Technology (S&T) or fields of Science and Engineering (S&E), used to classify R&D inputs and outcomes by fields of inquiry, namely, broad knowledge domains based primarily on the content of the R&D subject matter. Now explicitly includes R&D for the arts and humanities.
FOSE	Fields of Science and Engineering. See FORD, above.
FSS	Federal Support Survey, formally known as the NSF Survey of Federal Science and Engineering Support to Universities, Colleges, and Nonprofit Institutions.[11]
GBARD	Government Budget Allocations for R&D, formerly referred to as GBAORD, encompass all spending allocations met from sources of government revenue foreseen within the budget. Spending allocations by extra-budgetary government

[6] https://www.whitehouse.gov/sites/default/files/omb/assets/a11_current_year/s79.pdf

[7] http://www.nsf.gov/statistics/ffrdclist/#gennotes

[8] http://www.nsf.gov/statistics/srvyfedfunds/

[9] https://www.federallabs.org/

[10] http://www.oecd.org/sti/inno/Frascati-Manual.htm

[11] http://www.nsf.gov/statistics/srvyfedsupport/

	entities are only within the scope to the extent that their funds are allocated through the budgetary process.
GERD	Gross domestic Expenditure on R&D is total intramural expenditure on R&D performed in the national territory during a specific reference period. Includes current expenditures and capital expenditures.
GOVERD	Government Expenditure on R&D represents the component Gross domestic Expenditure on R&D (GERD) incurred by units belonging to the Government sector.
Grants to State and Local Governments	Includes the Federal Government's budget authority, obligations, and outlays to the 50 State governments and the District of Columbia, as well as county, municipal, township, school district, or special district governments as defined by the Bureau of the Census,[12] as well as Puerto Rico, the Virgin Islands, and other U.S. territories;[13] if Federal resources support State or local government operations or provisions of services to the public. These include, but are not limited to direct cash grants to State or local governmental units, to other public bodies established under State or local law, payments for grants-in-kind, payments to State and local governments for research and development that is an integral part of the State and local governments' provision of service to the general public (e.g., research on crime control financed from law enforcement assistance grants or on mental health associated with the provision of mental rehabilitation services).[14] These are a subset of extramural funds.
HERD	Higher education Expenditure on R&D represents the components of Gross domestic Expenditure on R&D (GERD) incurred by units belonging to the Higher education sector.
Internal R&D Funds	Amount of money spent on R&D that originate within the control of and are used for R&D at the discretion of a reporting unit. Do not include R&D funds received from other reporting units explicitly for intramural R&D.
Internal R&D Personnel	Persons employed by the reporting unit who contribute to the agency's intramural R&D activities.
Intramural R&D	All R&D performed within a reporting unit or sector of the economy during a specific reference period, regardless of the source of funds.
National Patterns of R&D	Report issued by the National Science Foundation on the overall R&D effort in the United States. Includes R&D expenditures by sector for performance, source of funds, and Type of R&D, R&D-to-GDP ratios, and international comparisons. Includes depreciation but not capital expenditures for R&D.[15]
Obligation	Amounts for orders placed, contracts awarded, services received, and similar transactions during a given period, regardless of when the funds were appropriated and when future payment of money is required.[16]
Outlay	Amounts for checks issued and cash payments made during a given period, regardless of when the funds were appropriated.[17]

---

[12] http://www.census.gov/govs/go/population_of_interest.html
[13] See OMB Circular A-11, Schedule C, Section 84.2(a) for more details:
https://www.whitehouse.gov/sites/default/files/omb/assets/a11_current_year/s84.pdf
[14] See OMB Circular A-11, Section 84.2(b) for more details:
https://www.whitehouse.gov/sites/default/files/omb/assets/a11_current_year/s84.pdf
[15] http://www.nsf.gov/statistics/natlpatterns/
[16] https://www.whitehouse.gov/sites/default/files/omb/assets/a11_current_year/s20.pdf
[17] https://www.whitehouse.gov/sites/default/files/omb/assets/a11_current_year/s20.pdf

Performers of R&D	Group, organization, institution, enterprise, or individual who conducts the R&D activities.
Personnel Costs	Salaries of both Federal employees who perform R&D as well as Federal employees who monitor or oversee R&D projects, both intramural and extramural.
PNPERD	Private Non-Profit Expenditure on R&D represents the component of GERD incurred by units belonging to the private non-profit sector.
R&D	Research and Experimental Development: comprise creative and systematic work undertaken in order to increase the stock of knowledge – including knowledge of humankind, culture, and society – and to devise new applications of available knowledge.
RD&D	Research, Development, and Demonstration – often broader than R&D alone. Demonstration is defined by the International Energy Agency (IEA) as the design, construction, and operation of a prototype of a technology at or near commercial scale with the purpose of providing technical, economic and environmental information to industrialists, financiers, regulators, and policy makers.[18] Only "technical demonstration" is to be included as part of R&D; "user demonstration" is not included as part of R&D.[19]
RDD&D	Research, Development, Demonstration, and Deployment. Deployment is the distribution or dissemination of products resulting from R&D. Deployment is not included as part of R&D.
RDT&E	Research, Development, Test, and Evaluation budget activities are broad categories reflecting different types of DoD activities related to science and technology.[20]
R&D Equipment	Major equipment used primarily for purposes of research and development. Includes acquisition or design and production of movable equipment, such as spectrometers, research satellites, detectors, and other instruments. For reporting purposes, at a minimum this should include programs devoted to the purchase or construction of R&D equipment, separate from R&D facilities themselves.[21]
R&D Facilities	Includes the acquisition, design, and construction of, or major repairs or alterations to, all physical facilities for use in R&D activities. Facilities include land, buildings, and fixed capital equipment, regardless of whether the facilities are to be used by the Government or by a private organization, and regardless of where title to the property may rest. Includes fixed facilities such as reactors, wind tunnels, and particle accelerators. Includes construction and rehabilitation of R&D facilities only. Exclude Other facility funding and movable R&D equipment.[22]
R&D Personnel	All persons engaged directly in R&D, whether they are employed by the reporting unit or are external contributors fully integrated into the statistical unit's R&D activities, as well as those providing direct services for the R&D activities (such as R&D managers, administrators, technicians, and clerical staff).
R&D Plant	The combined value of both R&D Facilities and R&D Equipment (as defined in OMB Circular A-11) reported to the *Survey of Federal Funds for Research and Development*.

[18] OECD. (2015), *Frascati Manual: Guidelines for Collecting and Reporting Data on Research and Experimental Development*, B.12.1

[19] Frascati Manual 2015, § 2.100-101.

[20] http://comptroller.defense.gov/portals/45/documents/fmr/current/02b/02b_05.pdf

[21] https://www.whitehouse.gov/sites/default/files/omb/assets/a11_current_year/s84.pdf

[22] https://www.whitehouse.gov/sites/default/files/omb/assets/a11_current_year/s84.pdf

SBIR	Small Business Innovation Research[23]
STTR	Small business Technology Transfer[24]
S&E	Science and Engineering
S&T	Science and Technology
Type of R&D	Refers to R&D activities of basic research, applied research, or experimental development. Also may be referred to as "R&D by type of work"; in previous iterations this may have been referenced as "character of work".
TRL	Technology Readiness Level – originally developed by NASA, TRLs are a measurement system used to assess the maturity level of a particular technology. Although similar, DoD and DoE have developed their own versions. Although elements of R&D by type of work can be found within the TRL, there is no formal crosswalk as some TRL levels may crosswalk to more than one R&D by type of work.[25]

Best Practices for the Identification and Reporting of R&D

The second goal in the FTAC-RDRS charter was to "share best practices associated with identifying and reporting data on Federal R&D spending data."[26] During two meetings in the fall of 2015, several FTAC-RDRS members from the National Science Foundation (NSF), National Aeronautics and Space Administration (NASA), Department of Energy (DOE), Department of Transportation (DOT), and the United States Geological Survey (USGS) made brief presentations describing their agencies' best practices. A summary of these best practices is provided below:

National Science Foundation (NSF)

NSF is one of a few agencies (along with the Smithsonian and the Nuclear Regulatory Commission (NRC)) whose topline R&D data reported to both OMB and the NSF Federal Funds Survey matches each year. All R&D reporting for the A-11 and the NSF surveys are done by staff in the budget office using a set of related reporting queries that access data from the financial and proposal/award management systems.

NSF captures character class information at the proposal level.[27] Character class information is entered by program staff as a series of percentages that indicate the nature of each project. NSF's proposal processing system's online help provides definitions of these character class categories, as stated in A-11. NSF has already linked its financial system to information in its grants system through the award ID. The ability to link the financial data to the proposal/award data allows analysts to access specific details such as whether or not the grant is for basic or applied research (a distinction which is applied to

---

[23] https://www.sbir.gov/about/about-sbir

[24] https://www.sbir.gov/about/about-sttr

[25] For NASA see: https://esto.nasa.gov/files/trl_definitions.pdf; for DoD see: http://www.acq.osd.mil/chieftechnologist/publications/docs/tra2011.pdf; and for DoE see: https://www.directives.doe.gov/directives-documents/400-series/0413.3-EGuide-04a

[26] See FTAC-RDRS Charter, pg. 2. https://community.max.gov/download/attachments/913670779/FTAC%20RDRS%202015%20charter%20-%20signed.pdf?version=1&modificationDate=1450292907483&api=v2

[27] Character class refers to reporting by agencies to OMB MAX Schedule C that distinguishes between investment and non-investment activities. R&D investments are one part of the character classification.

the award ID by the cognizant program officer at NSF during the review phase), as well as information about the place of performance, and type of performer. Field of R&D (or Field of Science) is based on the organization and appropriation funding the grant proposal.

National Aeronautics and Space Administration (NASA)

In the summer of 2013 NASA undertook a process to develop a R&D heuristic which has been used to compile data for both the A-11 and NSF R&D data calls. The process involved a series of meetings with the agency's mission directorates, budget staff, and the Office of the Chief Engineer. The result was a document that is a series of if-then statements that allows NASA to more accurately and consistently calculate R&D from non-R&D, Type of R&D (i.e., basic research, applied research, and experimental development), and how to objectively identify programs and projects by Fields of R&D. NASA staff revisits the heuristic each year as projects change and evolve. Although reporting is still a manual process, NASA is looking to develop more automation in the future based on the foundation that has been laid with the heuristic.

A copy of the NASA Heuristic has been posted to the FTAC-RDRS MAX page Documentation Archive for reference: https://community.max.gov/download/attachments/913670783/NASA%20R-D%20categorization%20hueristic%20PPBE%202016%20v1.docx?version=1&modificationDate=1447446900023&api=v2

Department of Energy (DOE)

During the past few years, DOE has been working on a process to better reconcile what are included as inputs into R&D reports by coordinating between Budget and Finance. The Budget office collects R&D information from program offices for the Budget Data Request (BDR), A-11, the NSF Federal Funds Survey, as well as for other requests by the American Association for the Advancement of Science, the Small Business Innovation Research and Small Business Technology Transfer programs, and for other Congressional justifications. DOE Finance Office collects fiscal year end data including depreciation, and the like necessary for reporting R&D activities in the agency's Annual Financial Report (note: OMB Circular A-136 requires agencies to report R&D activities in the annual financial reports and uses the same definition of R&D as is used by A-11 and the NSF Federal Funds Survey). Ideally the program offices should report the same programs and activities as R&D to both Budget and Finance, even though the request may be for different data (e.g., Budget Authority vs. expenditures on a modified accrual-basis). Budget and Finance staff noticed that, in some cases, they were getting different data from the same program offices for the two data calls. In some cases this was the result of different staff answering the two calls independently of each other, or reporting that did not account for changes in the program over time that would affect the results. In order to ensure consistent R&D inputs to both, Budget and Finance staff began to coordinate with each other on the inputs they were receiving from the program offices. In order to address these inconsistencies, Budget now includes references and copies of the submissions to Finance's calls and Finance includes references and copies of submissions to Budget's calls. Through this coordination and communication there has been a trend toward greater consistency in DOE's reporting of R&D.

Department of Transportation (DOT)

Eight of DoT's eleven bureaus have some R&D spending that should be reported to OMB and the NSF surveys. DoT has a process for tagging R&D, facilities, and technology so that they can provide amounts for each. R&D is determined at the project level, and is composed mostly of applied research and development activities. DoT does have separate definitions for R&D, facilities, and technology to help apply the distinctions at the project level. In addition, the DoT research, development, and technology (RD&T) tables (generated from an internal data call) are used to verify the R&D data in MAX Schedule C.

Department of the Interior, U.S. Geological Survey (USGS)

Referencing OMB Circular A-11 and the OECD Frascati Manual, USGS identified a need to better define how they should classify "monitoring" and "mapping". Based on guidance in the Frascati Manual, routine mapping and monitoring activities would not be considered R&D, however the development of new processes or mechanisms for mapping and monitoring activities would be consistent with both the international guidance and Circular A-11. As a result of these changes there were some shifts in the R&D and non-R&D percentages for a number of programs in the 2016 President's Budget Request. This has resulted in more accurate reporting of USGS R&D activities to both OMB and the NSF-NCSES surveys.

A copy of the USGS presentation has been documented on the FTAC-RDRS MAX page Documentation Archive: https://community.max.gov/download/attachments/913670783/RD%20at%20USGS%20-%20for%20FTAC.pptx?api=v2

Guidance for Reporting R&D Boundary Issues

This section provides standard guidance for navigating the boundary issues that commonly occur when determining what to report for any Government R&D data call. The following documentation includes scenarios, provided by FTAC-RDRS members based on agency experience where the intended meaning was not obvious and where additional guidance would be helpful.

The third goal in the FTAC-RDRS charter was to "develop a reporting model to help improve accuracy and reduce reporting burden associated with Federal agency R&D spending data for OMB and NCSES data calls."[28] This reporting model would aid Federal agencies in responding to OMB and NSF data calls on their R&D funding activities. The combination of compiled definitions, glossary, best practices, as well as the guidance and R&D schema discussed below are the foundation of the R&D reporting model.

Although all data calls use the same definition of R&D, there is an element of interpretation required to determine what activities should be included as R&D and which should not. The FTAC-RDRS devoted substantial time and effort to discussion of topics that were considered on the boundary of R&D, and many agency participants noted the need for guidance to help them standardize responses across data

---

[28] See FTAC-RDRS Charter, pg. 2.
https://community.max.gov/download/attachments/913670779/FTAC%20RDRS%202015%20charter%20-%20signed.pdf?version=1&modificationDate=1450292907483&api=v2

calls. Based on the international standards for reporting R&D used by both OMB and NSF, the co-chairs have drafted guidance to address each scenario.

As the FTAC-RDRS is succeeded by future working groups, this guidance document will be modified and updated accordingly by future groups to aid future respondents and to address new areas where interpretation of the R&D reporting requirements require additional guidance.

This section is organized around the major components of R&D. Namely, definitional – whether or not activities should be considered R&D; Type of R&D (i.e., basic research, applied research, and experimental development); Performance of R&D – intramural or extramural; Place of Performance; Field of R&D; and other conditions not elsewhere classified.

Definitional – is it R&D?

Testing and evaluation

All three OMB Circular A-11 definitions of R&D[29] require additional guidance to provide clarity to reporting agencies. Circular A-11 separates basic and applied research into unique categories but then lumps all other R&D categories, as traditionally used by most DOD[30] and many non-DoD agencies, into one "Development" category. It may be useful to incorporate some of the language from these definitions to more precisely define each category, such as test and evaluation.

- Should test and evaluation be reported in R&D? For example, our agency has certain categories that are similar to the DOD 6.4-6.7 categories; should we report these as R&D?

- Is there a difference between Science and Technology (S&T) and the three R&D categories listed in A-11?

- Where does "development" stop; is it before or after test and evaluation or later fielding?

Guidance:
Generally speaking, a good rule to help determine where and when experimental development (including testing and evaluation work) stops and is no longer part of R&D as defined would be prototyping. If the agency is still working on the development of a prototype, those activities should be reported as R&D. Experimental development should not be confused with pre-

---

[29] Basic Research: Systematic study directed toward fuller knowledge or understanding of the fundamental aspects of phenomena and of observable facts without specific applications towards processes or products in mind. Basic research, however, may include activities with broad applications in mind. Applied Research: Systematic study to gain knowledge or understanding necessary to determine the means by which a recognized and specific need may be met. Development: Systematic application of knowledge or understanding, directed toward the production of useful materials, devices, and systems or methods, including design, development, and improvement of prototypes and new processes to meet specific requirements. (International guidance on R&D definitions have recently changed, which may be incorporated into future revisions of A-s11 guidance.)

[30] DoD RDT&E Categories 6.1: Basic research, 6.2 Applied research, 6.3 Advanced Technology Development, 6.4 Advanced component development and prototypes, 6.5 System development and demonstration, 6.6 Research, Development, Test, and Evaluation (RDT&E) Management Support, 6.7 Operational Systems Development. 6.1-6.3 combined are referred to as "S&T"; 6.4-6.7 combined are referred to as Major Systems Development.

production development. Identifying the precise cut-off between the two requires engineering judgment as to when the element of novelty ends and the work changes to routine product development. However, in some instances of later-stage development activities, problems may still arise during these late-stage tests and evaluation of these tests and new experimental development may be needed; this "feedback R&D" should still be reported as R&D.[31] If most experts agree that significant feedback R&D is likely to occur, it is an indication that the project is still prototyping, and in an experimental (vs. pre-production) stage of development. Most late-stage pre-production development as described above would not be R&D.[32]

For most agencies, R&D is a subset of a broader concept of Science and Technology (S&T) that encompasses R&D, demonstration, deployment, STEM education, operational observation and data collection and other such activities utilizing science and technology. For DOD, Science and Technology (S&T) is specifically defined as a subset of R&D ("6.1" through "6.3" categories).

Each R&D project consists of a set of R&D activities, is organized and managed for a specific purpose, and has its own objectives and expected outcomes (even at the lowest level of formal activity).

Development and technology/demonstration

The definitions for Research and Development (R&D) come from OMB Circular A-11. However, it is standard practice for the Department to include "Research, Development & *Technology*" activities in budget reporting exercises.

- In OMB A-11, **"Development"** is defined as *"systematic application of knowledge or understanding, directed toward the production of useful materials, devices, and systems or methods, including design, development, and improvement of prototypes and new processes to meet specific requirements."*
- The Department defines **"Technology"** as *"Demonstration projects and other related activities associated with research and development activities."*

Clearly, these two terms are very similar, making it difficult for analysts to separate the two in budget reporting documents. The "Technology" category is also used to cover "research deployment" activities including technology transfer, training, education, and technical assistance.

Questions:
- Should the Department continue to report "Technology" separate from "Development"?
- Is there still value in reporting "Technology" if no other Federal agency does so?
- Can [the Department] obtain a more accurate delineation of the two terms?
- Should the "Technology" term be revised or substituted for something else?

---

[31] Frascati §2.36

[32] For an activity to be classified as R&D it must meet all 5 core criteria: Novel (directed towards new findings), Creative (focused on original, not obvious, concepts and hypotheses), Uncertain (outcome, cost, time allocation, etc. are not known a priori), Systematic (planned and budgeted), and Transferable and/or reproducible (results could be transferred or reproduced either internally to the organization or shared broadly). If the activity meets all 5 of these criteria it should be considered an R&D activity. In most cases, R&D activities can be grouped to form R&D projects. See Frascati §2.13-2.20

- How do other Federal agencies address the definition and reporting of research deployment/implementation activities?

Guidance:

Generally speaking, some activities associated with "Technology" as described here may be included as part of R&D activities. There are two concepts of demonstration projects in measuring R&D: "user demonstration," which takes place when a prototype is operated at or near full scale in a realistic environment to aid the formulation of policy or the promotion of its use, which is not R&D; and "technical demonstration" (including the development of demonstration projects and demonstration models), which, because it is an integral part of an R&D project, is an R&D activity. In this case "Technology" is synonymous with technical demonstration projects related to an R&D activity and should be included.[33] With reference to its broad use in the management of large research projects, "technology demonstration" is seen as a step in the process of evaluating, ex-ante or ex-post, the implementation of new technologies and generally means the activity carried out to show to potential investors and customers the expected potentiality of a technology under development. In this respect, the use of this concept and term is not recommended for use in association with the R&D concept, unless a clear role of a "technical demonstration" activity in an R&D project could have been identified.

The FTAC will not comment on inside-Department reporting categories. If a separate "Technology" category has value for the Department, then the Department may wish to continue reporting on it even if there is no external audience for it. But it may be useful for the Department to consider an alternate term such as "Technology Demonstration." The Department of Energy, for example, has distinct development, demonstration, and deployment categories for its science and technology activities.

Classification of Mapping/Monitoring Use Case

Some clarification should be provided for when mapping and monitoring activities should be considered R&D. NSF has defined "Research" as "Systematic study directed toward a more complete scientific knowledge or understanding of the subject studied." Other groups define research in a similar manner.

As such, one could argue that any mapping or measurement of the earth system is considered research because it increases knowledge about the system. But this interpretation does not pass a common-sense test. Routine mapping of previously mapped areas and measuring temperature and water levels for routine reporting and operational predictions does not meet the expectation for more complete understanding that is part of the definition of research, even though some routine observations could have ancillary uses in research. On the other hand, mapping unexplored areas of the ocean and measuring greenhouse gases are typically considered research.

---

[33] Frascati §2.100

Some guidelines for distinguishing which mapping/monitoring is research would be helpful. For instance, one approach might be to consider the purpose for which the mapping/monitoring is conducted. If the primary purpose is to produce routine products and services (e.g., updated map quads, tomorrow's weather forecast), the work is likely to *not* be research. However, if the primary purpose is to support research activities, then it is more likely to be research—even though the collection of measurements could use standard instruments applied in a routine manner. In other words, the purpose of the work may be more important in these cases for determining whether the work is research than the type of activity performed (i.e., mapping and monitoring). Other approaches to classification could also be considered.

Guidance:
Generally speaking *routine* mapping and monitoring activities are not considered R&D as they lack some (viz., novelty and creativity) of the five core criteria of R&D activities, even if the collection is done by R&D personnel. However, the purpose of the data collection effort may help determine if the activities should be reported as R&D. For example, if data collection is part of routine mapping or monitoring efforts and are for broad dissemination or standard map production or as part of an agency's mission to manage a public trust (e.g., clean water/air), then those activities should not be classified as R&D. On the other hand, if data collection is for analysis or use in testing a particular hypothesis, or if the data collection is for investigation to gain or enhance a general understanding of an area or to acquire new knowledge then these activities should be reported as R&D.

Impact assessments, program evaluation, and the like

The current definition of R&D excludes *routine* program monitoring and evaluation. The definition does not specifically address Impact Assessments. Given that [our] agency invests in Impact Assessments, which utilizes research methodologies, to evaluate program and performance, it will be helpful if a decision is made to either include or exclude the Impact Assessment from the R&D definition. Currently we do not capture investments in Impact Assessments as part of our agency's research portfolio. If an Impact Assessment is considered an R&D activity, then should it be classified as applied research, or experimental development?

Guidance:
Generally speaking, impact assessments and program evaluations are not considered R&D, unless they are evaluations of an R&D project/program itself.

R&D efforts may underpin the decision-making process within government and other institutions. For example, an agency may have a dedicated team actively involved in carrying out original analyses on an ad hoc or even formalized basis. Considering that the definition of research includes activities that increase the stock of knowledge in people, culture, and society, such activities may be classified as applied research, as they are intended to generate new knowledge (more than develop new products and services), but they have a direct purpose in mind (supporting decision-making to advance an Agency's mission). However, this is not always the case, and not all impact assessment or evidence-building efforts associated with policy or programmatic advice can be correctly described as R&D. It is relevant to consider whether the

impact assessment process is routine, and can be completed by collecting a standard set of observations, or whether it requires a high level of expertise and creative thinking.
See the box below for additional guidance:

Is the evaluation measuring the program performance of…	Non-R&D activities?	Or R&D activities?
Is the program evaluation approach novel? Will you be creating new methods that will inform future evaluation efforts?	That is R&D – it is generally applied research to support decision-making	That is R&D – including, for example, the Science of Science Policy
Is it a routine or standardized program evaluation, such as those completed to meet a long-term Congressional mandate?	That is not R&D.	That is R&D – because it is an administrative cost for managing an R&D project.

Fellowships and training

Our agency supports a large variety of education activities including traineeships, fellowships, etc. MAX A-11 Character Class Data requires that these activities be reported separately from R&D. The NSF Survey appears to want these included if they are R&D related. Because a large part of our mission is R&D, many of our education activities are science or R&D related. How do we reconcile these reporting requirements?

Guidance:
With regard to R&D, generally all education and training of personnel whether that training is in the natural sciences, engineering, medicine, social sciences, the arts, and the like should be excluded from R&D reporting. However, it is particularly difficult to establish the boundaries between education and training activities and R&D for doctoral students. Parts of their curricula for studies are highly structured, involving, for instance, study schemes, set courses and compulsory laboratory work. Here, the teacher transmits knowledge and provides training in research methods. Students typically attend compulsory courses, study the literature on the subject and learn research methodology. These activities, and the support they receive to undertake them, do not fulfil the criterion of novelty specified in the definition of R&D.

In addition, students are also expected to prove their competence by undertaking relatively independent study, usually containing the elements of novelty required for R&D projects and presenting their results. These activities, often supported through research training grants or tuition remission arrangement for students working on research, should be classified as R&D. In addition to R&D performed within the framework of postgraduate education courses, it is possible for PhD or even Masters students to be engaged in other R&D projects. Further, such students are often attached to, or directly employed by, the institution in which they study and have contracts or similar engagements that oblige them to teach at lower levels or to perform other activities, and may receive teaching assistantships allowing them to continue their studies and to do research. Such support is not R&D.[34]

Agencies are encouraged to develop a well-informed estimate of education activities versus R&D activities for the participants in any STEM Education program or fellowship, and use that percentage to inform their reports of the character of work supported by that program.

Type of R&D

a. Our agency funds a wide range of scientific projects from basic science to field-based clinical trials. For the past 12 years, the Institute of Education Sciences (IES) has required that applicants select one of five research goals around which they build a research project. These research goals are Exploration, Development and Innovation, Efficacy and Replication, Effectiveness Evaluations, and Measurement. We use these goals to structure their response to this data call. We consider all research funded under Exploration to fit the category of Basic Research, all research under Development and Innovation to fit the category of Experimental Development, and research funded under the remaining three categories to fit under the category of Applied Research.

However, other projects are more difficult to identify. This is particularly true for larger R&D projects, such as National R&D Centers or Research Networks where teams are asked to carry out research that includes multiple different types of research. We typically estimate what proportion of their research fits into each of the categories (using the goal structure described above); however, research projects are not required to report out how many dollars are associated with different types of research activities within a larger grant. To what degree of specificity should we be breaking projects apart?

Guidance:
Since the agency's own administrative data systems do not provide additional detail of specificity for Type of R&D among larger, more complex R&D programs, some estimation is needed. Generally speaking we would defer to the program manager to make the appropriate designation as to how much may be considered basic research, applied research, or experimental development. Ideally, a program manager might identify specific R&D projects within the larger program and assign those largely by type of work and split the dollars accordingly for the sake of reporting. However, we recognize that the ability to identify specific tasks of projects by Type of R&D may not be available and in that case, developing *a well-informed percentage split of activities by Type of R&D* may be acceptable, although it is strongly recommended that the logic of the decision be documented as part of

---

[34] Frascati §9.37-9.41

an agency's own internal controls and that these logic models/heuristics be reviewed each year and updated accordingly as projects and programs evolve.

b. We define Applied Research (APR) as systematic study to gain knowledge or understanding necessary to determine the means by which a recognized and specific need may be met. APR is undertaken either to determine possible uses for the findings of basic research or to determine new methods or ways of achieving specific and predetermined objectives. The results of applied research are intended primarily to be valid for possible applications to products, operations, methods or systems. Some examples of projects we classify as APR are:

- Investigations undertaken in an effort to distinguish between antibodies for various diseases is applied research.
- A clinical trial to determine efficacy and effectiveness in priority health areas.
- Researchers plant experimental crops where they alter the spacing and alignment of the plants to reduce the spread of disease while ensuring the optimum arrangement for maximum yield.
- Researchers select for traits associated with water use efficiency in greenhouse studies to develop drought tolerant wheat and rice varieties.

Development Research (DVR) is defined as the systematic application of knowledge or understanding, directed toward the production of useful materials, devices, and systems or methods including design, development, and improvement of prototypes and new processes to meet specific requirements. Some examples of DVR projects include:

- Researchers create a tool for gene editing by using knowledge of how enzymes edit DNA.
- Implementation science or operations research to guide introduction and uptake of proven interventions.
- Researchers use a portable, fluorescence-based technology to detect and mitigate aflatoxin contamination in maize.
- Research into how hermetic triple bags used for chemical-free storage of cowpea could be used to store other crops.

Given that the differences are nuanced, it becomes difficult for Agency staff in the field to distinguish between these categories. This leads to potential for improper coding and errors that are carried forward. For the purpose of our sponsored research distinguishing between these categories may not be useful. Having a single category "Research" to report will ensure that investments are recorded more accurately. However, in lieu of that option what guidance or "markers" might we provide staff to help make these distinctions with more clarity? Is there a meaningful difference between these types of projects or can we classify all as APR or DVR instead of trying to split them between these two categories?

Guidance:
Generally speaking the definitions used by the agency for APR and DVR best map to the standard definitions of Applied Research and Experimental Development, respectively – as defined in Circular A-11, A-136, and the NSF Federal Funds Survey. Nonetheless it is possible that some borderline projects could be late-stage applied research or early-stage experimental development and that having certain "markers" to help make the determination could be useful.

Agencies may wish to compile a logic model, similar to the one used in the above example, that documents real-world examples of the agency's R&D portfolio that can be used to distinguish projects among the categories of basic research, applied research, experimental development, and non-R&D. In the above example, it would mean clinical trials are applied research, gene-editing tool development is experimental development. There will always be boundary cases that will not fall cleanly or clearly into one category. Using a well-reasoned logic model to make the best determination that also documents the agency's rationale for the classification are important to ensure consistency with future reporting of similar projects and would be part of an agency's own internal controls for external reporting purposes.

Performers of R&D

a. Using economic support funds, [the Department] awards a grant to the National Academy of Sciences to solicit and fund bilateral research between the United States and country X. The desired outcome from the grant is collaborative research and exchange of students (graduate) between the labs. While the NAS reviews and solicits proposals using a standard merit-based methodology, the science and resultant publications are only captured for output purposes. Is this R&D and if it is, should it be captured? (Note: this type of project is pretty common; it is the same model used for the Cooperative Threat Reduction following the fall of the Soviet Union.)

Guidance:
It appears that R&D is taking place, although it may be mixed with other non-R&D activities (e.g., exchange of graduate students/education). In this case it seems prudent to clearly identify the roles and responsibilities of each of the parties with regard to the research projects, and how they are funded and by whom. If there are agency labs and staff involved in the research these should be considered intramural. Although the exchange students brought in to work on the research may be external to the agency they are still supporting an intramural project and should be reported as part of intramural R&D. It appears that NAS is functioning as an outsourced research administrator. Thus, if the NAS, is serving as a facilitator/administrator of projects that are deemed intramural then the funds to NAS to administer the research programs are themselves intramural.

b. The Department has Federal employees who use space provided by state academic institutions. Departmental employees use state supported offices, laboratories, and greenhouses. This space and these facilities are crucial to the Department's science mission. In many cases, the Department does not reimburse the institution for use of this space. No future Federal funds appear to be available to support facilities and infrastructure for these locations. Should we report the Research and Development costs that could be associated with using these offices, laboratories, greenhouses, etc. because of how important they are to performing the USDA mission? Should these be reported separately from the actual expenditure of Federal funds?

Guidance:
Expenditures for R&D facilities, major equipment, labs (collectively, R&D Plant) should always be reported separately from expenditures for the conduct of R&D. If the agency has budget authority, obligations, or outlays associated with R&D Plant – either support to state or academic research space, or their own facilities, this should be reported as R&D plant. If Federal funds are not used to support research or research space, then the Federal agency should not report these costs.[35]

Place of Performance

a. Most of the Department's research grant awards involve research being carried out in a location other than that of the prime. Sometimes this occurs when an award involves a prime and sub; other times this occurs when the prime is carrying out research activities in school districts outside of the prime's physical location, and sometimes those activities may be occurring across the country. While it is relatively straightforward to break budgets apart by prime and sub, it is next to impossible, given our current budget reporting lines for research grants, to pinpoint exact dollar amounts spent to collect data in each school district or state where data is being collected. At what level of specificity do we need to report the places of performance data?

Guidance:
The most requested data sets on Federal R&D funding are related to place of performance. Although we recognize that expenditures by specific place of performance may not be available, agencies should at least report to the finest level of detail available. Generally speaking, priority should be given to the organizational structure of an activity than to the literal location of where the activity takes place in classifying "intramural" R&D that takes place outside of the physical location of the reporting unit. For example, a grant to a university research who is part of University A, and located in region X, may occasionally perform short-term field work (as part of unit A's R&D project) physically in region Y. Unless there are countervailing reasons (e.g., financing arrangements with another unit in another region), all of the intramural expenditures for that R&D may be reported for region X as opposed to region Y.[36] Generally speaking this guidance is also consistent with DATA Act implementation guidance for place of performance.

b. Our Agency funds research both within the United States and internationally. The research support is provided by Central funds from Washington Bureaus as well as Missions in the field. While it may be possible to capture place of performance at the Country-level for activities funded by the Agency's Mission offices in that particular country, it is challenging to disaggregate this information from Regional Missions that have responsibility for multiple countries in their region and from a research portfolio of the Agency's Washington Bureaus

---

[35] This guidance is for funder-based surveys. R&D Performer-based surveys would recommend the calculation of an amount that represents the user costs of the facilities. See Frascati §4.35
[36] Frascati §4.164

that may fund an activity internationally. This is also compounded by Prime vs. Sub issues. The FTAC discussions on this topic seemed to convey a desire to capture detailed information. Given the diverse location the Agency sponsors activities in, it may not be cost effective to capture this data accurately. We have programs where an entity has been awarded a contract to administer a research program. While the entity is located in the United States, it funds research activities across the globe. Are there recommendations for how best to report place of performance given these constraints on the data availability without resulting in additional burden on the Agency?

Guidance:
The most requested data sets on Federal R&D funding are related to place of performance, including by foreign nation. Although we recognize that expenditures by specific place of performance may not be available, agencies should at least report to the finest level of detail available. For funds that are awarded to or administered by U.S. institutions, but that clearly support R&D performed outside the United States, a best-guess estimate of R&D by performing nation may be the most cost-effective approximation available.

Field of Research and Development (FORD), formerly Fields of Science and Engineering

We do not currently disaggregate the research data by subfield of science. Currently all the information is contained in one category, as majority of fields of research supported by the Agency fit under "Life Science". This is not a concern for us. An annual report on the Agency's R&D, issued by the Agency's Research Coordinator will have information on research investments that are disaggregated by the sector and major programs. Given that we have no internal information on subfields available to query for these data calls, how might we best estimate subfields to fulfill the requirement without adding additional burden? Is it possible to report the top-line field total and impute the subfields to rake to the total?

Guidance:
In the event that detailed subfield data are not available at the agency level but the appropriate project manager can reasonably estimate a split based on the nuance of the work involved in the projects, then agencies should ask project managers to review periodically their portfolios to obtain subfield splits. However, in the event that a knowledgeable researcher or project manager cannot identify these subfields, agencies may provide estimated percentage splits across the subfields that add to the total. However, as a matter of internal controls for external financial reporting, these decisions should be documented and periodically reviewed for changes based on the evolution of the research programs. In addition, an agency logic model/heuristic, once developed and maintained, would ensure consistency of projects classified as R&D between the A-11, A-136, and NSF data calls for R&D.

SBIR, STTR, and FLC contributions

We have recently significantly increased the amounts of the R&D activities we report with the inclusion of supporting administrative activities. These costs include such as program direction, safeguards and securities (including cyber security) and infrastructure projects. These supporting costs budgets are often tight and our programs are concerned about further reductions. Should these supporting administrative costs for R&D be included when calculating SBIR/STTR or to a lesser extent the Federal Labs Consortium contributions?

Guidance:
Generally speaking, administrative costs associated with an agency's own intramural R&D as well as extramural contracted R&D programs should be included as intramural (when all of the R&D consists of extramural grants, none of those administrative costs are R&D).[37] Therefore these costs are outside the scope of the baseline for the SBIR and STTR assessment. The inclusion of safeguards and security in some cases may be a necessary cost of conducting R&D. If these safeguards and security measures are contracted out to support an intramural facility these costs are still intramural R&D and not subject to the baseline for the SBIR assessment.

Administrative costs

a. The reporting requirements for R&D under the A-11 and the NSF-Federal Funds Survey call for the inclusion of administrative costs. Our agency's respective financial and management systems do not have separately identifiable information on the administrative costs (salaries and expenses) associated with R&D contracts, and grants. How should we estimate these costs to meet the reporting requirement? Do we need to go to each office that manages R&D work and identify each staff member involved, how much of their time per year is dedicated to these activities and cross-that with salary and overhead data to compile an estimate? We do not have the resources to identify costs that way. Is there an easier way to provide this information that would still provide a reasonable estimate for the data call?

Guidance:
Given the constraints imposed by the agencies' own financial and management systems, it is unreasonable to ask agencies to account for each and every employee's time and salary devoted to R&D administration. A well-informed estimate that accounts for differences in staff time and salaries may be used to report these data. However, to ensure consistency of what is included as R&D expenditures across different data calls, it is prudent for agencies to document the methodology for estimating these data and review them periodically as projects evolve, end, and new ones begin.

b. The reporting requirements for R&D under the A-11 and the NSF-Federal Funds Survey call for the inclusion of administrative costs. We have one account-code that is used across the agency to fund all administrative activities. However, this account does not have separate records for administration of R&D vs. administration of procurement contracts. If we

---

[37] Frascati §8.35

include the costs from that account it will include a lot more than is just for administering R&D. Should we leave it out and not report any administrative costs or can we apply a ratio to the account to estimate the amount that is *likely* attributable to R&D work?

Guidance:
Administrative costs are as much a part of doing R&D as is the salaries of scientists and support staff and they should be included in an agency's R&D reporting. In this case, even though all administrative costs are taken from a single agency-wide account the entire account should not be included as it includes administrative costs for non-R&D projects as well. Ideally, universal identifiers between agency financial and management systems should allow one to develop more reliable estimates of R&D costs from this type of agency-wide account. Agencies should, in the meantime, develop a well-informed percentage split of R&D projects which would be supported by the agency-wide account and use that to pro-rate the administrative costs to R&D. These should be documented as part of an agency's own internal controls for the record to ensure consistency across R&D data calls.

c. How to capture administrative costs: clarification needed on the definitions of administrative costs – i.e., overhead vs. direct office administration vs. indirect grant/procurement support.

Guidance:
Administrative costs for R&D include, but is not limited to, staff salaries, overhead, direct office administration time and costs dedicated to R&D projects, as well as indirect and direct procurement support.

d. How to capture Facilities costs: When is O&M considered Non R&D? And if O&M is performed on a facility that does some R&D, is it acceptable to prorate the costs?

Guidance:
Operation and Maintenance (O&M) should be included as part of the costs of R&D for an agencies own intramural R&D activities or O&M costs associated with extramural contracts for R&D. It should be noted, as per the international standards to simplify reporting, that any O&M costs that are solely dedicated to extramural R&D grants should not be counted as R&D.

O&M costs associated with an R&D facility should also be included as R&D facility costs. However, if an intramural facility, which is not primarily an R&D facility, but does have some R&D space and functions within, it is acceptable to prorate the O&M costs to those facilities assuming the agency can identify the facility space separately from all other non-R&D space.[38]

---

[38] Frascati §4.36-4.37

R&D nested within other programs/activities

a. Within our accounting system, R&D is not captured as the primary program area or object classification in most cases. For instance, a piece or suite of equipment may be purchased which is charged off to our digital forensics program and designated as information technology hardware from an object classification standpoint but then ultimately is used in the development and fielding of a new evidence extraction capability. This makes identification of research and development extremely difficult without the data call responder having first-hand knowledge of the situation or manually reviewing purchases across the organization, which isn't practical due to the size of the organization and sheer volume of transactions. This brings into play questions such as could a model for capturing certain costs be developed, should a reporting threshold amount be included for each category of R&D, or should a standard assumed percentage of R&D for certain types of programs be developed?

   Guidance:
   For programs in which R&D is not the primary activity, then agencies should develop a model for assigning a well-informed percentage of program expenses, perhaps based on a one-time audit or manual review of sample programs, which could be used to estimate a program's R&D funding. These percentages should be documented as part of an agency's internal controls for the record to ensure consistency across R&D data calls, and should be updated periodically.

b. Our Department funds the Regional Education Laboratories, which conduct applied research; seeks out and widely disseminates findings from high-quality research through live events and a wide range of media; and provides training, coaching, and technical support for applying research to education improvement. Using task-based statements of work, we can typically estimate budget amounts associated with applied research. Is this method of estimation acceptable?

   Guidance:
   Yes, this is acceptable. It is recommended that the methodology for reviewing and classifying information based on these statements of work are documented as part of an R&D logic model.

R&D and reimbursable agreements

We often receive funding from other government agencies to support collaboration on the development of investigative tools. For instance, one of our intelligence community partners may provide us funding under the Economy Act to support the development and potential fielding of an information gathering tool for which both agencies can benefit and utilize as we address our agency specific missions. This brings into mind two areas of needed guidance. First, should we (the receiving agency) capture and report those costs or should each agency report the costs independently? Second, should we, as an agency with mission responsibilities in both the intelligence community and the domestic criminal community, even report intelligence community assets?

Guidance:
Generally speaking, if your agency will be conducting R&D under a reimbursable agreement or inter-agency agreement you should not include those funds received from another federal agency in your own agency's reporting. The agency that is transferring the funds to your agency, for your agency to use for intramural R&D activities should report these funds as R&D expenditures on their response.

Intelligence community-supported R&D should, as with all other Federal R&D, be reported to the R&D surveys by the agency providing funding for the R&D. Agencies may need to seek guidance on a case-by-case basis from OMB and NSF NCSES on how to disclose intelligence-community and other classified R&D activities.

R&D and non-appropriated funds

The FBI has the authority to establish and collect fees to process fingerprint identification records and name checks for non-criminal justice, non-law enforcement employment and licensing purposes and for certain employees of private sector contractors with classified government contracts. The fee structure under this authority covers two distinct areas, a dollar for dollar cost recovery portion to fund salaries and other expenses incurred to provide these services, and, an automation surcharge portion to support the automation of fingerprint identification and criminal justice information services and associated costs. This second, surcharge authority often contains elements of research and development as next generation systems and software are developed and implemented. Should these research and development costs be captured as this funding is not received through appropriations but from the private sector via statutory authority? If so, where and how should they be reported? Would the same be true for other types of funds that are appropriated but not annually such as multiple or no-year type appropriations?

Guidance:
Any R&D funded from fees as described above should be reported as R&D expenditures from the account that collects the fees. While a portion of the fees collected may be transferred to other accounts within the department for the purposes of R&D they should be reported from the account of origin.

As per OMB Circular A-11 Reimbursable means an obligation financed by offsetting collections credited to an expenditure account in payment for goods and services provided by that account. (See section 83.5.) If your account includes reimbursable obligations (see section 20.5), show the obligations financed by reimbursements separately from direct obligations.

R&D facilities/plant

We are building a new facility where R&D will be performed. There are plenty of construction costs required for the facility that are not core to the R&D space. For example, the construction and connection of a new sewer line from the facility to the local sewer authority. We do not have a way to separate the major capital costs for those that are core to the facility (e.g., construction of the lab space or addition of fixed equipment) from other costs (e.g., sewerage or adjoining office space). Should we report the construction of these additional costs for the sewer or adjoining office space as part of the facility or should we separate, and if we should separate the costs how should we estimate for them when no detailed costs are available from accounting system?

Guidance:
If the facility in question is *primarily* going to be used for R&D and there is no way to separate these construction costs then they should be included as part of the costs of construction of the R&D facility. If, on the other hand, the facility is mostly office space with some marginal amount of space for R&D then these should be excluded.

Standardized R&D Reporting Schema

The R&D Reporting Model consists of the Definitions and Glossary of Terms associated with R&D reporting, examples of existing best practices in identifying and reporting R&D, guidance on how to address specific questions about R&D, and the R&D Reporting Schema.

What is it?

The purpose of the R&D Reporting Schema is to describe the structure of the existing R&D reporting requirements, in order to support the development of automated reporting methods. The schema is the underlying organizational structure/outline of all of necessary variables, for each data call, and how they relate to one another. It addresses all three of the major R&D data calls (viz., OMB's MAX A-11, NSF's Federal Funds and Federal Support Surveys, and submissions to Treasury for the Government-wide Financial Report as per OMB Circular A-136).

How is it useful?

After an agency has identified which programs are and are not R&D and classified activities by Type of R&D, Type of Performer, Field of Research, etc. (using the guidance documentation or through the development of agency-specific logic models or heuristics) the schema provides the framework for organizing that information for reporting purposes.

The schema provides a model for how variables roll-up (or down) to the top-line R&D totals. What rolls-up to the R&D total for OMB should be consistent with what rolls-up to the R&D total for the NSF-NCSES *Federal Funds Survey*. Similarly, what is reported as applied research for one data call is consistent with other data calls. Although the definitions of applied research, for example, are consistent across data calls, *where* the data are reported on the various data calls may differ – the schema provides information on where the data should be reported in relation to other variables. Use of the schema is dependent on development of standard definitions, reporting guidance, and agency-specific heuristics/logic models. The schema provides a model for how we can, in the future, report data in machine readable format. The schema also provides the core framework needed for establishing an R&D reporting exchange on the National Information Exchange Model (NIEM) which could be used to facilitate automated reporting through structured data files.

The schema is organized in a series of MS Excel Workbooks, one for each data call.

Reporting is not just for R&D totals, but all nested variables and how they roll-up to the top-line R&D totals. The R&D Reporting Schema can be found on the FTAC-RDRS MAX page Documentation Archive: https://community.max.gov/download/attachments/913670783/Standard%20RD%20Schema%20for%20OMB%20NCSES%20Treasury%20Data%20Calls%20v4.3.xlsx?api=v2

Updates and revisions to the schema will be managed by OMB and NSF-NCSES in cooperation with the proposed Federal R&D Reporting Community-of-Practice.

FTAC-RDRS Summary: Accomplishments, Recommendation, and Next Steps

The fourth and final goal in the FTAC-RDRS charter was to "propose a long-term governance structure for continued management and oversight of the government-wide R&D reporting standard."[39] The FTAC-RDRS completed the goals in the charter, but there is a continued need to maintain and build upon the efforts to date.

Accomplishments

The FTAC-RDRS co-chairs certify the major accomplishments, with intended follow-up items in italics, were:

1. Collected existing R&D definitions and reporting requirements for R&D spending – *These definitions and terms and references to R&D reporting requirements will continue to be available on MAX as a reference for Federal agencies.*
2. Shared agency best practices for identifying and reporting data on Federal R&D spending – *Documentation of FTAC sharing of agency best practices will be available as a reference for Federal agencies.*
3. Developed a reporting model to help improve accuracy, quality, and consistency of Federal agency R&D spending data and reduce reporting burden associated with OMB and NSF-NCSES annual data calls – *A glossary of terms, agency requests for guidance and FTAC guidance, use cases, R&D schema, heuristics, and other material comprising a reporting model will be available as a reference or guidance for Federal agencies.*
4. Proposed a long-term governance structure for continued management and oversight of the government-wide R&D reporting standard - *There will be a set of recommendations for Federal R&D agencies to consider for possible action by Committee on Science, OMB, and/or OSTP. There will be an informal interagency working group to carry forward the work of the FTAC-RDRS, and*

---

[39] See FTAC-RDRS Charter, pg. 2.
https://community.max.gov/download/attachments/913670779/FTAC%20RDRS%202015%20charter%20-%20signed.pdf?version=1&modificationDate=1450292907483&api=v2

an Executive Office of the President (EOP)-led consulting group to work with individual agencies on applying the FTAC-RDRS reporting model to improve specific agencies' R&D reporting.

The FTAC proposes the following recommendations be approved by the Committee on Science:

It is recommended that Federal agencies that fund R&D create consistent documentation of which programs and activities are R&D, which Types of R&D are supported, and of R&D performer classifications (extramural versus intramural). In order to simplify internal analysis and external reporting, it is recommended that agencies map the Type of R&D work to the program and budget activities currently tracked in each agency's budget formulation and execution systems.

It is recommended that an informal team of agency and EOP experts on R&D reporting and definitions be available to work with agencies to refine and improve their documentation of which budget activities are R&D, and the type of work of those R&D activities. The primary goal of this team is to increase the alignment of R&D reporting with government-wide and international standards as much as possible.

Agencies that have been early adopters of R&D Reporting best practices are encouraged to work with agencies that are beginning to document the R&D content of their programs and activities.

Early adopters and the NSF-NCSES are encouraged to work with the EOP to develop more automated R&D reporting methods in order to reduce the agency burden, and increase the accuracy of the R&D reporting process. The FTAC Co-chairs expect the R&D Reporting Schema will enable pilots of automated R&D reporting methods. Once such pilots have become successful, the examples will be collected as an R&D reporting toolkit to be shared with other departments and agencies. However, use of the R&D Reporting Schema is not recommended until an agency-specific R&D reporting heuristic has been developed.

Next Steps:

There will be an informal Community of Practice to carry forward the work of the FTAC-RDRS, and an EOP-led consulting group to work with individual agencies on applying the FTAC-RDRS reporting model to improve specific agencies' R&D reporting.
The Community of Practice for R&D Reporting will have several functions:
1) Early adopters and R&D Reporting subject matter experts will be available to mentor and work with agencies beginning the process of documenting and improving the characterization of the R&D work within their programs and activities.
2) Review of the existing glossary, guidance, and boundary cases will occur at least once every two years, but more often if the Community of Practice determines that there is an urgent need.
3) The Community of Practice will also serve as a resource for the R&D data collection staff of OMB and NSF, providing feedback and new ideas on ways to improve the R&D reporting process.
4) Early adopters of R&D reporting best practices, NSF NCSES, and various staff involved in the R&D reporting process from throughout the EOP will work together to develop an R&D Reporting

Toolkit.   This toolkit will include examples of automated R&D reporting methods in order to reduce the agency burden and increase the accuracy of the R&D reporting process.

Once a toolkit for improved R&D reporting is developed, the Community of Practice will work on broadly sharing the R&D Reporting toolkit, possibly through the NSTC or an interagency memo.

Appendix A – FTAC-RDRS Participants

Co-Chairs

Kei Koizumi, Associate Director for Federal R&D, Office of Science and Technology Policy

Sue Romans, Budget Director, National Aeronautics and Space Administration

John Jankowski, Program Director R&D Statistics, National Center for Science and Engineering Statistics, National Science Foundation

OMB Liaison

Celinda A. Marsh

Executive Secretary

Christopher V. Pece

Department of Agriculture

Richard Derksen

Paul Heisey

Department of Commerce

Richard Cavanagh (NIST)

Jason Boehm (NIST)

Steven Fine (NOAA)

Marian Westley (NOAA)

Vicki Schwantes (NOAA)

Department of Education

Elizabeth Albro

Sue Betka

Department of Energy

Marcos Conzales-Harsha

Mark Joseph

Beverly Kipe

Department of Health and Human Services

Carol Linden (FDA)

Rakesh Raghuwanshi (FDA)

James Onken (NIH)

William Duval (NIH)

Brian Haugen (NIH)

Department of Homeland Security
Richard Williams

Department of the Interior
Cynthia Lodge (USGS)

Anne Barrett (USGS)

Mary Cantrell (USGS)

Department of Justice
Patrick Devall (FBI)

Department of State
Samuel Howerton

Danyal Petersen

Department of Transportation
Alasdair Cain

Patrick Sandy

Melissa Stanley

Department of the Treasury
Raymond Sendejas

Beverly Babers

National Aeronautics and Space Administration
Amy Kaminiski

National Science Foundation
Elizabeth Velo

NSF-National Center for Science and Engineering Statistics Liaison
Michael Yamaner

U.S. Agency for International Development
Michael Curtis

Pallaoor Sundareshwar

Peter Boyle